THE
TRUTH
ABOUT
SPACE

THE TRUTH ABOUT SPACE

H.L.A. FRASER

Library of Congress Control Number:		2024916573
ISBN:	Hardcover	979-8-3694-9729-6
	Softcover	979-8-3694-9728-9
	eBook	979-8-3694-9727-2

Print information available on the last page.

Rev. date: 09/30/2024

To order additional copies of this book, contact:
Xlibris
AU TFN: 1 800 844 927 (Toll Free inside Australia)
AU Local: (02) 8310 8187 (+61 2 8310 8187 from outside Australia)
www.Xlibris.com.au
Orders@Xlibris.com.au
846771

Preface

There are two theories in quantum physics called string theory and M-theory. They are closely related to one another, and you could describe M-theory as an extension of string theory.

These two theories play a key role in this book, so for the benefit of those who have never heard of them, I'll give a brief rundown of what they're about and the reason why they are important here.

At around the time that Einstein was developing his theory of relativity in the early twentieth century, which deals with the large-scale structure of the universe, a fellow German physicist, Max Planck, was also developing the theory of quantum mechanics, which deals with subatomic structures.

These two theories were obviously quite different from each other. So the idea was to try to unify them into one theory that would become known as the theory of everything.

The theory of everything also aimed to unify the four known energy forces and, most importantly for this narrative, discover the most elementary particles in the universe from which all other particles have evolved. And for the past century this quest has basically become the Holy Grail of theoretical physics.

String theory arrived on the scene somewhere in the 1970's, and after a rocky start it has now become the most popular theory in modern physics.

The success of string theory lies in its ability to accurately predict the nature of fundamental particles when they exist at an incredibly small size known as the Planck length. This is a size that is measured as 10^{-33} cm at 10^{-43} seconds. To bring that into perspective, one analogy is to say that if an atom was the size of the solar system, then a particle at the Planck length would be the size of an atom.

String theory is named after the particles of energy that exist at the Planck length because they are seen not as particles but as tiny strings of energy that vibrate at various frequencies with each frequency giving rise to a different particle.

One of the mysteries of string theory that had researchers baffled was that the theory was describing five different types of string that could each describe the particles of the universe in different ways. These five types are called the type I, type IIa, type IIb, SO(32), and E(8) xE(8).

But in 1997 physicist Edward Witten discovered that these five types could be unified into one theory which became known as M-theory.

The reason why these two theories are significant here is because of the number of dimensions that they each require. String theory shows that when strings exist at the Planck length, they have to be existing in a space that has a total number of ten dimensions. And M-theory states that there has to be one more dimension, called the eleventh dimension, which co-exists with the ten dimensions of the strings but is uniquely different in its own way. It's these eleven dimensions that are central to the theme of this book.

Chapter 1

1987.

Recently I watched the movie *Around The World In 80 Days,* based on the novel by Jules Verne. The version I watched was the 2004 version starring Steve Coogan and Jackie Chan. At one point in the movie, the lead character, Philias Fogg, played by Steve Coogan, made a statement that I immediately picked up on. What he said was "All truths are born from facts", and that struck a chord with me straight away because it reminded me of the situation I found myself in in 1987.

Back then I was twenty-five years old and living with a problem that I had developed half a lifetime earlier. I was a living, breathing, practicing alcoholic. And it was this affliction that was the direct cause of a particularly devastating catastrophe that occurred one night in July of that year. It was a catastrophe that left two good friends in hospital, both with broken bones, and I knew full well that it was my own drunken stupidity that had been the cause of this event.

The four days following that catastrophe were without doubt the worst four days I have ever lived in my life. The overwhelming sense of grief, remorse, shame, self-loathing, and a whole host of other soul-crushing emotions, were all doing their damndest to make sure they were going to teach me a lesson I was never going to forget. And they all kept at it mercilessly and continuously throughout the entire duration of those four days.

By the fifth day I began to recover from the shock. My friends were still in hospital and doing as well as could be expected under the circumstances. But I was now left with a major personal problem that I knew had to be dealt with urgently.

Over the years my drinking had been the cause of many problems. But usually they were issues that I could at least live with, like employment issues or driver's licence issues. But this latest calamity that I had once again been responsible for was on a different level altogether. I had put people's lives at risk, and I knew I had pushed my reckless drinking behaviour to its absolute limits. It was time to call in the big guns of the drug and alcohol rehabilitation community. And for me that meant two things: Alcoholics Anonymous and a non-residential rehabilitation facility known as Holyoake, both of which I was already well familiar with.

This is where the statement made by Philias Fogg at the beginning of this narrative starts to become relevant.

Over the preceding four years I had been intermittently attending both A.A. meetings and Holyoake sessions in a continuous attempt to overcome my addiction problems. But time and again these attempts failed as I found it all too easy to just return to the old habits and vainly hope that everything was going to be alright.

This time, however, there was something new. The experience of those four "soul-crushing" days had done something to my mind. They had left me with a steely determination to do something positive and constructive with my life. I wanted so desperately to deal with my alcohol problem once and for all, and now I was thoroughly determined to succeed.

But although my drinking behaviour was obviously the most demanding problem I had to face, it wasn't the only one. Because in returning to A.A. and Holyoake there was another problem I knew I'd been avoiding that I was going to have to face up to and deal with. And this problem was all to do with Alcoholics Anonymous and one of the recommendations it made in its program.

A.A. is based on a twelve-step program which participants are encouraged to observe and put into practice to the best of their ability. And for me it was the advice given in step three that I had a real problem with.

Step three recommended that we "turned our will and our lives over to the care of God as we understood him".

I knew what this recommendation meant because it was based on the Christian world view that God is our creator and therefore he gives our lives meaning and purpose.

But for me it wasn't going to be that simple, because turning my life over to the care of God would mean by necessity accepting the Christian decree that God's word is the truth. But if God's word is the truth, then that would mean that God is a giant, invisible human being who built the universe and everything in it in six days.

And he would have to be a giant human being because he built mankind according to his own likeness, in his own image, as it says in *Genesis*, the first book of the Bible.

The problem I had with all this was that I was living in an era where our ever-refined scientific knowledge was telling me that life on planet Earth had evolved over millions of years, on a planet that was 4.6 billion years old, in a universe that was three times older than the planet. And by then we had amassed plenty of evidence to show that this account of our history was essentially true.

So what this meant for me was that if our knowledge of evolution and the history of the universe was true, then the biblical account of Genesis wasn't true. But that didn't necessarily mean that God didn't exist; it just meant that we got it wrong the first time.

So the question I was left with was the question of whether or not God actually existed. And that was the question that I needed an answer to.

But after those four days of living hell that I had just put myself through, I wasn't going to be content with just finding any answer.

What I wanted was the absolute truth. If God wasn't a giant, invisible human being, who didn't build the universe in six days, then who or what was he? And did he even exist at all? If I was going to achieve any success with A.A.'s step three then this was the problem that needed to be addressed.

Because I wasn't about to hand my life over to someone whose very existence was questionable. And with my newfound determination I wasn't going to bypass this problem either. This time I was going to deal with it.

I wanted to know the truth about whether God existed or not. And I wanted that truth to be based on facts that I could know and prove to myself to be true with absolute certainty. Nothing less was going to be sufficient, because having purpose and meaning in my life was the one thing I was lacking. In some ways it would've been one of the underlying reasons for why I became an alcoholic in the first place. But I couldn't just turn my life over to the care of an entity that did more to confuse me than it did to enlighten me.

I needed the truth. And I needed the facts to be able to prove that truth. The problem I had was the question of how to establish those facts. But as luck would have it, it was going to be Holyoake that was going to provide me with the first two facts that were to become the foundation for everything that was to follow.

So by July 1987 I had once again booked myself into Holyoake for the umpteenth time. By now I was well familiar with the Holyoake routine, and it was this familiarity that allowed me to focus on and critically analyze the lectures I was learning. As a result of this it was two lectures in particular that sent me into a state of deep thought. One was a lecture on the patterns of human behaviour and the other was a lecture on the human belief system. It was from analyzing these two lectures that I was able to establish my first two facts.

The first fact was derived from the lecture on the patterns of human behaviour. I deduced from this lecture that any naturally occurring pattern would be something I knew I could trust because naturally occurring patterns are not influenced in any way by the human mind. We have no power to decide what a naturally occurring pattern will do. We can only observe it.

From the second lecture, regarding the human belief system, I concluded from this lecture that the one thing I know I cannot trust is my own belief system because my belief system allows me to believe whatever I want to believe without having the necessity of having to prove whether that belief is actually true or not.

So now I had established my first two facts which were:

Fact 1: There is one thing I know I can trust.

Fact 2: There is one thing I know I cannot trust.

The one thing I know I can trust is any naturally occurring pattern, and the one thing I know I cannot trust is my own belief system.

These two facts now became the foundation of truth from which I could build my knowledge and understanding. I've since learned that in scientific circles they would be what's known as "first principles" which basically means a self-evident truth from which to start.

Now that I had these two facts established it was time to analyze what they meant. And this is what led to the next mind-numbing discovery that was to establish the next two facts I was going to be able to prove.

As I sat and contemplated what it meant to have a belief system I knew I could no longer trust, something strange began happening to my mind. This can be likened to what happens when you have a bathtub full of water and you remove the drain plug. You see all the water in the bath steadily drain away until there is nothing left. And this is pretty much what happened to the contents of my mind. I realized that if I have a belief system I can no longer trust then how could I be sure about anything I presumed to know if it was all based on assumptions that I couldn't prove to be true? This left me with the realization that there was nothing I could prove to myself at all.

But eventually the one thing I was forced to accept that I knew I could prove to be true was the fact that I existed as a conscious, thinking being. I knew that that had to be true because otherwise I would not have been able to consciously draw that conclusion.

Then secondary to that conclusion I realized that if I exist then I must be existing in space because if I didn't exist in space then I wouldn't be able to move around freely.

So these conclusions now gave me the next two facts that I could prove to myself:

Fact 3: I know for a fact that I exist as a conscious, thinking being.

Fact 4: I know for a fact that I exist in space.

And now with these two facts established it was time to return to fact number one.

Fact one proved that the only thing I knew I could trust was any naturally occurring pattern.

So with the only two things I could prove to myself being the fact that I exist and the fact that space exists, the next question was, are there any naturally occurring patterns involved either in the physiological structure of myself or in the physical structure of space? Because if there were any such patterns, I knew above all else that they would be patterns I knew I would be able to trust.

So looking for these patterns in my own physiological structure wasn't going to be too difficult because I exist in a physical form with a clearly defined shape. So locating any patterns in that shape wouldn't be too hard to detect.

But space was an altogether different thing because although it's something we all know we exist in, it's also something our senses are completely immune to. And that's when I realized that if I was going to be dealing with the structure of space then I was going to have to do it by dealing with the structure of the dimensions of space.

The Pattern of The Physiological Structure of The Human Body.

When I went looking for patterns in the human body, I started by focusing on the brain because that's the place where consciousness lives. As I considered this I realized that the 'brain' and the 'mind' are really two different definitions of the same thing. From the external point of view it can be seen as a brain, but from the internal point of view it can be seen as a mind.

So what you see from the external point of view is the brain as part of the head and head as part of the body.

What you see from the internal point of view is the mind which is connected to its external environment via the five senses of the body.

There are four senses of the head, which are sight, sound, taste and smell, and one sense that covers the entire body including the four head senses, and that is the sense of touch. So the bodily sense of touch is part of the four senses of the head, and all five senses are part of the internal sixth sense, which is the mind.

So what I had discovered was a naturally occurring pattern that showed two diametrically opposite points of view. From the external point of view, it showed how the brain is part of the head, and the head is part of the body. And from the internal point of view, it showed the opposite: The bodily sense of touch is part of the four head senses and all five senses are part of the mind.

In summary the pattern was showing me the internal point of view and the external point of view both together at the same time.

Chapter 2

Dimensions.

When I turned my attention to the dimensions I was starting this investigation with a clean slate. I had just emptied my mind of all its accumulated clutter, and I was dealing with a subject I knew virtually nothing about. So the first thing I needed to do was to formulate an idea of how I was going to go looking for patterns in the way dimensions were structured.

I decided that if I started with the two most basic models I could think of, which were the cube and the sphere, I could use these models to determine what, if any, patterns might arise from the differences in their geometry.

The next thing I needed to do was to find out what a dimension actually was. I knew that space was three-dimensional, and time was the fourth dimension, but that was about the limit of my knowledge. That being the case, I did what I always do whenever I get stuck on the definition of a word; I looked it up in my copy of *The Concise Oxford English Dictionary.*

The dictionary started off by listing the three dimensions of length, breadth and thickness. These are the three linear dimensions that define Three-dimensional space. But I decided that the three names the dictionary had used weren't appropriate, so I opted to use the names of width, height and length instead, which is what they will continue to be called throughout the rest of this book.

Width has a direction that goes from left to right, Height goes up and down, and length goes forward and backward. And these are the three dimensions that are used when we say that space is three-dimensional.

Along with these first three dimensions the dictionary also included the dimensions of area and volume in the same list. It then went on to explain that a line is one-dimensional, though not a dimension itself, and that a point was considered to be non-dimensional.

As I considered these definitions, I was beginning to see something about them that didn't seem to make sense, because the dictionary had included area and volume as measurable dimensions but not point and line.

This didn't make sense to me because in my mind point, line, area and volume were all part of the same geometry. So it didn't make sense to call just two of them dimensions but not the other two. As far as I was concerned, if you were going to call two of them dimensions then why not call all four of them dimensions? That was what made sense to me and so that's what I did.

But that wasn't the only problem I had with the dictionary's definition, specifically in relation to the value of a point. If I started with the dictionary's description of a point as being non-dimensional, then if I drew a single line from that point then that line would become the first dimension. So what would happen then when I reversed the procedure and removed the line? That would leave behind the point, which was considered to be non-dimensional. Or to give it a more precise mathematical definition, zero-dimensional.

But what would happen if I continued with my deduction and removed the point as well? That would leave behind nothing. And nothing was also measured as being zero-dimensional. So now I had two separate entities, nothing and point, that were both sharing the same dimensional value of zero.

When I realized this, that's when I decided to introduce zero as a measurable dimension to represent the proper value of nothing.

This meant that I now had a list of five dimensions which were zero, point, line, area and volume.

But there was still one more dimension to add to this list which was the dimension of time. And this was due to Einstein's theory of relativity which introduced time as an extra dimension of space and giving rise to the term 'spacetime'.

When I added zero as a measurable dimension, I justified it in my mind by comparing it to the decimal digit of the same name and value. I figured that if you can have a decimal digit that records the value of nothing then why not a spatial dimension as well?

At the time I thought it was going to be the only comparison I was going to make between the decimal digits and the dimensions, but as it turned out, it was only the beginning.

So now that I had this new list of six dimensions, these were the dimensions I was going to be able to use to build my two models of the cube and the sphere. And because with the cube, the three dimensions of line, area and volume all followed the same directions as the three linear dimensions of width, height and length, my assumption was that all I had done was to take the standard definition of four-dimensional spacetime and added the extra two dimensions of zero and point, bringing the total number of dimensions to six. This assumption turned out to be wrong. But it was going to be an entire decade before I was going to discover this error and the reason why I made it.

In the meantime, it didn't matter anyway because the six dimensions I had now listed were the six dimensions that I needed to build my two models.

At this point I would've liked to have included the two images of the cube and the sphere to demonstrate the pattern that emerged, but that ain't gonna happen because I don't know how to do artwork on a computer and there's no one to help me. So instead I'll just describe the images and leave it to the reader to use their imagination, which shouldn't be too difficult. It isn't exactly rocket science.

So with the cube I started with a straight horizontal line leading away from the point. This represented the first two dimensions of point and line. The next dimension was area, which was shown as a square extending above the line. Then the dimension of volume was extended behind the square to complete the shape of the cube.

Time was included as a wrap-around dimension just to show that it was included as part of the geometry.

With the sphere the dimension of line started at the point then extended as a circle that ended back at the point. The dimension of area then extended inside the line to represent the surface of the sphere. Then the dimension of volume extended inside the dimension of area to complete the structure. Time was added as a wrap-around dimension as it was with the cube.

Everything that existed outside of the cube and the sphere was the nothing that was represented by zero dimension.

What these two models were showing me was that with the cube, the line extended away from point then area extended outside the line and volume extended outside area.

The sphere was showing me the opposite. The line extended back to the point and then area extended inside the line and volume extended inside the area.

The models were showing me two diametrically opposite points of view. Between them they were showing me the internal point of view and the external point of view both together at the same time.

And this result was the exact same result that I arrived at with the pattern of the physiological structure of the human body.

Once I had these two patterns established, I knew straight away that this was an important discovery for human knowledge, and I knew this for two fundamental reasons. The first reason was because of the rigorously enforced and thoroughly scrutinized sequence of logical procedures I had used in order to arrive at their observation. And the second reason was that although they were two completely separate patterns, they were both showing me the exact same result. Both patterns were showing me the internal point of view and the external point of view both together at the same time.

The problem I now had was trying to find an explanation for the reason why these two patterns in particular were so fundamentally important. And that's where I came to a dead end. No matter how I tried I simply couldn't find any way of explaining their significance. I was totally stumped, and that's the way it remained for an entire decade.

Chapter 3

<u>String Theory.</u>

It wasn't until September 1997 that I came across an article in *New Scientist* that caught my attention because it was dealing with the question of why we live in a universe with three dimensions of space and one of time. It mentioned that this question was relevant because of a theory in quantum physics called string theory. It went on to say that string theory required as many as 26 dimensions.

This article was my introduction to string theory as well as quantum physics in general. Prior to this I had not even been aware that string theory existed, even though it had already been around for at least a couple of decades.

I was obviously interested in this article but not enthusiastic. The reason being that this was an article that was dealing with theories in quantum physics, of all things. But for me, I had completed year 10 at high school when I was fifteen years old. And that marked the end of my illustrious career in the education system. On top of that, string theory was a theory, I surmised, that would be using some of the most complex mathematical systems on the planet: the kind of stuff that takes years and years of intense university study to master.

But here I was, with my two simple models of a cube and a sphere, models that were so basic even a child would have no problem understanding them.

So the article was interesting. But the idea that string theory could in any way be even remotely connected to my own work on dimensions was just laughable, and I knew it! So the article was put to one side, ignored but not forgotten.

But as time rolled by over the following weeks, I slowly came round to the idea that after ten whole years with no new discoveries, it was this article alone that had been the one and only item that had provided even the slightest connection to the work I was involved with, even though that connection was only due to the existence of one particular word in the article; 'dimensions'.

I knew it was a bit of a long shot for someone like me to be dealing with theories that were at the very heart of quantum physics, but by this stage I'd decided that I had nothing to lose anyway. It wouldn't do any harm if I made the attempt to investigate string theory further, because you never know, maybe physicists understood something about dimensions that I was unaware of that could be of some assistance to my own investigation.

So with that in mind I headed off to my local library in search of any books that might be available on the subject. And once again, as luck would have it, I found the one book on the shelves that was going to turn out to be specifically the very book that I needed.

It was a book called *Hyperspace,* written by Michio Kaku, an eminent New York physicist. And it was this book that contained the one crucial piece of information to help me realize what it was that I had so critically overlooked in 1987.

Hyperspace was published in 1994. By then physicists had established that the tiny strings of energy that gives string theory its name vibrated in a specific number of higher dimensions of space. And for string theory there were two different versions. First there was the string theory version which vibrated in 26 dimensions of space. Then there was the superstring version which vibrated in ten dimensions.

I didn't know it at the time, but years later I was to discover that the 26-dimensional model of the theory wasn't dealing with the number of dimensions. What the number 26 was actually referring to were 26 specific properties of the ten dimensions. But nobody knew this at the time, and I will explain it in more detail in a later chapter.

What I learnt from the book is that strings of energy vibrate at a size known as the Planck length, which is billions of times smaller than the size of an atom. And what happens at this size is that they vibrate in a space consisting of a total of ten dimensions. But so long as they exist in this state, they remain violently unstable. And it's this instability that forces the ten dimensions to rupture and split into one group of six dimensions which collapses, and one group of four dimensions which can then expand into the four-dimensional spacetime of our universe.

It was this description of the ten dimensions breaking up into two separate groups that became the main focus of my attention because what it showed me for the first time was that dimensions could be separated into two different groups: specifically one group of six and one group of four.

When I took this description and applied it to the two models of the cube and the sphere the connection was immediate. I realized that what I had done in 1987 was to use the same six dimensions to describe both models, which revealed the pattern of the two diametrically opposite points of view. But what the dimensions of string theory were showing me was that there could now be two separate groups of dimensions. There could be one group of four dimensions to describe the internal point of view, and one group of six dimensions to describe the external point of view.

Suddenly the penny dropped, and I realized what it was that I couldn't figure out in 1987. My assumption at the time had been that the three dimensions of the cube - line, area and volume - were representative of the three linear dimensions of width, height and length because they all followed the same directions. But now I realized that the four dimensions of width, height, length and time were the dimensions that could be used to describe the internal structure of space, and the six dimensions of the sphere - zero, point, line, area, volume and time - were the dimensions that could be used to describe the external structure of space. Two groups of four and six dimensions making a total of ten, exactly the same as the dimensions that string theory had predicted.

This was an incredible discovery, and I was speechless.

Could it be that the string theorists and myself had been using two completely separate and isolated mathematical systems that both ended up arriving at the same ten-dimensional structures of space? Because if that wasn't the case then it meant that this was nothing more than a really weird and utterly bizarre coincidence.

But how was I going to find the answer to that question? It was pretty obvious to me with my high school education that I wasn't about to go delving into the mathematics of string theory. So one thing I did decide to do was to rush off a letter to Michio Kaku briefly outlining what I had discovered, but I never got a response. That isn't surprising because the contents of that letter would've sounded to Kaku like the rantings and ravings of a complete lunatic.

So in the end I could see that the only thing I could really do was to just persevere, keep reading the magazine articles and the books until I came across the next clue that was going to help me with my quest.

Missing The Point.

After I had separated the dimensions into their two separate groups, I noticed a curious anomaly concerning the four internal dimensions. Because if they were now existing as a separate group then shouldn't there be a point at their centre from which they could all expand into their individual directions? This is what seemed to make sense but there were a couple of reasons why it couldn't be allowed. The first reason was that the point had already been allocated to the group of six external dimensions and it couldn't be duplicated. And the second reason was that adding a point would turn the four-dimensional structure into a five-dimensional structure, and that didn't make sense at all. So it was an annoying anomaly, but I had to let it go because there was nothing I could do about it.

M-Theory.

In 1998 I bought another book by Michio Kaku called *Beyond Einstein* and co-written by Jennifer Thompson. It was first published in 1987 but the copy that I purchased was printed in 1997, and it contained some revisions and a new afterword at the end of the book.

The afterword was all about the latest discovery in string theory called M-theory. This theory showed how there could be one more dimension added to the ten dimensions of string theory. It was called, not surprisingly, the eleventh dimension and although it was part of the overall group of dimensions it differed from the ten-dimensional strings because it existed more as a membrane of energy rather than a string of energy.

My immediate assumption was that this new dimension would have to be another linear dimension, just like all the other linear dimensions. And this is what threw me because there was no way I could make it fit into the ten-dimensional model I had developed. It just didn't fit in at all and once again I was totally stumped.

In 1999 I spent some time in Adelaide, South Australia and whilst I was there I visited two physicists who were working at the University of Adelaide. I spoke with each of them only briefly, but it was here that one of them recommended another important book for me to read that dealt with the subject of string theory. It was called

The Elegant Universe by Brian Greene, another eminent New York physicist. This book was published in 1999.

I went to the university library, located a copy and began reading it straight away. It took a few days but even after I'd finished reading it, I was still no closer to understanding what the eleventh dimension meant.

Five years later, in November 2004, the three-part series of *The Elegant Universe* was shown on television in three weekly episodes. And when I found out that it was going to be on, I knew I was in for a special treat, especially when it came to the third episode which bore the ominous title '*Welcome To The Eleventh Dimension*'.

I've recently been able to review this episode on YouTube, and now I can't spot what it was that triggered my thinking when I first saw it in 2004. But there was obviously something Brian Greene said that made me realize that the eleventh dimension didn't necessarily have to be a linear dimension. It could just as easily be a point dimension.

That was it! The eleventh dimension was the point from which the four internal dimensions of the universe could expand into four-dimensional spacetime.

I couldn't believe it. I had actually solved the mystery of the missing internal point by identifying it as the eleventh dimension.

With the eleventh dimension firmly established, by January 2005 I had all that was necessary to write out the complete geometric structure of eleven-dimensional space based on the definitions provided by Michio Kaku in *Hyperspace* and *Beyond Einstein.*

I deduced from this geometry that what happens when a ten-dimensional structure ruptures and splits into two separate four- and six-dimensional structures, this is the mechanism that forces the eleven-dimensional membrane into existence. Because it is the eleventh dimension that is needed to keep the two groups permanently separated from each other whilst maintaining their connectedness.

This is what allows the six-dimensional external structure to remain collapsed whilst the four-dimensional internal structure can start to expand and cause a four-dimensional universe to exist.

This function of the eleventh dimension, being identified as the internal point, turned out to be a crucially important discovery because it was knowledge that was vital in helping me make the final connection that was to establish the proof I had been searching for.

It wasn't until two months later, in March 2005, that I had one of those lightbulb moments.

I was with some friends at the time, and at one point during the evening I suddenly blurted out that I had just had a revelation.

This 'revelation' was the realization that the ten decimal digits have actually got an eleventh digit which is called 'decimal point'. And this point serves the same function to digits as the internal point does for the dimensions. In both cases the point is used to separate two opposites.

With the decimal digits the decimal point is used to separate whole numbers from fractional numbers. And with spatial dimensions internal point is used to separate internal dimensions from external dimensions.

Once I had made this connection it didn't take long for me to recognize that each of the eleven decimal digits has a value that is equal to each of the eleven dimensions.

Over March and April of that year, after almost two decades, I was finally able to write down the mathematical proofs for the geometric structure of eleven-dimensional space.

Chapter 4

The Mathematical proofs For The Geometric Structure of Eleven-Dimensional Space.

To begin with there are two elementary statements that can be made regarding the decimal digits and the spatial dimensions.

Statement 1.

Decimal Digits

There are ten decimal digits plus one eleventh digit which is called decimal point. One of the ten digits is called zero. Of the remaining nine, five of them are odd digits and four of them are even digits.

Statement 2.

Spatial Dimensions

There are ten spatial dimensions plus one eleventh dimension which is called internal point. One of the ten dimensions is called zero. Of the remaining nine, five of them are external dimensions and four of them are internal dimensions.

Decimal point is equal to internal point and zero digit is equal to zero dimension. The five odd digits are equal to the value of the five external dimensions and the four even digits are equal to the value of the four internal dimensions.

The graph below shows how each of the digits are equal to each of the dimensions.

INTERNAL DIMENSIONS

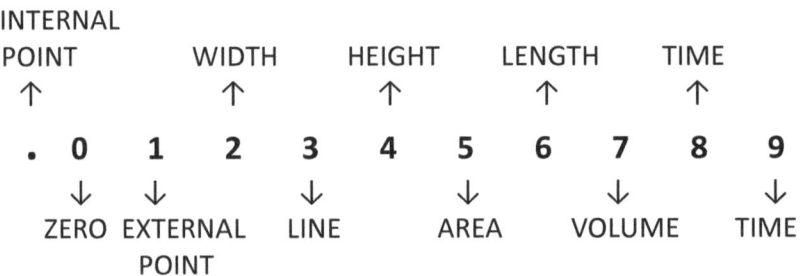

EXTERNAL DIMENSIONS

With the internal dimensions, width has two lines of direction that go to left and right. This gives it a numerical value of two. Height adds another two directions of up and down for a value of four. Length adds two more directions, forward and backward for a value of six, then time adds the two directions of past and future for a value of eight.

With the external dimensions, external point is the first dimension with a position only but no lines of direction.

So that singular position gives it a value of one. Line is the next external dimension with two lines of direction. So the one position of point plus the two directions of line makes a value of three. Area adds another two lines of direction for a total of five, and volume adds two more directions to make seven. Finally, time adds the two directions of past and future for a total of nine.

Time And Time Again.

If you're wondering why there are two dimensions of time it's because the four-dimensional internal structure of space is describing the structure of 'spacetime', which includes time as the fourth dimension due to Einstein's theory of relativity.

The six-dimensional external structure is describing the same spacetime from a diametrically opposite point of view.

So if it's spacetime from the internal point of view then it is also spacetime from the external point of view.

So both points of view are going to require their own dimension of time. That's why there are two dimensions of time.

The Geometry of A Basketball.

Imagine you have a basketball in front of you.

Now place your mind inside the ball at a point at the centre. From that point you can expand outwards into the three linear dimensions of width, height and length. Time is also included in this geometry because time is one of the dimensions of spacetime. So for the sake of simplicity just run time along the same line as length to represent the directions of past and future. This gives us the four-dimensional internal structure of spacetime.

Now place your mind on the outside of the basketball.

Start with a point at the base of the ball. This marks the first external dimension of point. Next draw a line from the point that goes around the circumference and back to the point. This is the second dimension of line. Draw a second line from the point round the circumference and perpendicular to the first line. This line represents the third dimension of area. Draw one more line straight down from the north pole to the south pole and this represents the dimension of volume.

Once again time has to be included so just draw it as an extra line around the circumference.

Everything that is outside of the basketball is the nothing that is represented by zero dimension.

This now gives us the six-dimensional external point of view of spacetime.

In this model it will be noted that the internal dimension of height and the external dimension of volume both follow the same geometric line. Despite this they both remain as two separate dimensions with two separate lines of direction: one goes up and down and the other goes from the south pole to the north pole and from the north pole to the south pole.

Also, these two dimensions along with the two internal dimensions of width and length, and the two external dimensions of line and area, all of these dimensions maintain the rule of remaining perpendicular to one another.

So how do you use the geometry of a ten-dimensional basketball to explain the geometry of a ten-dimensional string? Which is a different shape entirely.

A simple explanation would be to use a common party balloon as an analogy.

If you blow up the balloon it will inflate to a shape that is roughly similar to the basketball. Not a perfect sphere like the ball but close enough. This now represents the structure of the ten-dimensional sphere.

What you do next is deflate the balloon entirely then grab it by both ends and stretch it out as far as it will go without snapping, so it is stretched out into the shape of a long thin straight line.

This procedure has changed the shape of the balloon entirely from that of a sphere to that of a straight line. But it is still the same balloon and therefore still the same ten-dimensional structure.

Of course this isn't what actually happens to a string, it's just an analogy used to show how it is possible to use the geometry of a ten-dimensional sphere to explain the geometry of a ten-dimensional string.

The point is that the basketball is used to explain the structure of the space, not the shape of the space.

Position And Direction.

In the chapter on symmetry in *Beyond Einstein,* Kaku and Thompson discuss the mystery of why there are three numbers in particular that keep 'cropping up' in the superstring theory. They mention that even the mathematicians can't figure out the reason why they are so predominant.

The three numbers they were referring to are the numbers **8**, **10** and **26**.

I will now explain why those three numbers keep cropping up.

1. Zero
2. Point
3. Line
4. Area
5. Volume
6. Time
7. Width
8. Height

9. Length

10. Time

This is a list of the ten dimensions in order. Starting with the six external dimensions then the four internal dimensions.

In this list, the first two dimensions of Zero and Point are the two that have a position only but no lines of direction.

The remaining eight dimensions each have one position and two lines of direction.

So there are a total of **8** dimensions that have direction and a total of **10** dimensions that have position.

The eight linear dimensions each have two lines of direction making a total of sixteen individual directions. And sixteen directions plus ten positions makes a total of **26** positions and directions.

So the numbers **8**, **10** and **26** are important to superstring theory because they are the three numbers that are the totals of all the positions and directions of the ten dimensions.

The Anatomy of A Coin.

Since I live in Australia I'll be using a coin of Australian currency for this demonstration. But the reader can just substitute it for a coin of their own national currency.

A twenty-cent coin is a flat metal disc with two sides printed on it. One side is heads and the other side is tails. If it didn't have these two sides imprinted into its surface it would just be a metal disc with a scrap metal value of probably something less than one cent. But because it has these two sides it is given a fixed value of twenty cents that everybody in the nation, without exception, agrees upon.

What this coin does is it provides a solid object that can be used to represent a fixed mathematical value, which in this case is the value of twenty cents, that can be universally accepted by one and all.

Now you can flip this coin many times and each time it will land sometimes on heads and sometimes on tails. These are two different sides with two completely different images on them. But regardless of which side the coin lands on, the one thing that remains constant throughout the exercise is the coin's value. Whether it's heads or tails it still retains its ascribed value of twenty cents.

Now imagine that you have a full set of eleven coins. But these coins are not going to have heads and tails printed on them. Instead, each coin will have a single decimal digit on one side and a spatial dimension of the equivalent value on the other. Below is the list of these coins with the digits and dimensions that each coin represents.

Decimal Digits Side	Spatial Dimensions Side.
Decimal Point.	Internal Point.
Zero.	Zero.
One.	External Point.
Two.	Width.
Three.	Line.
Four.	Height.
Five.	Area.

Six.   Length.
Seven.   Volume.
Eight.   Internal Time.
Nine.   External Time.

So for example the coin with the number 2 printed on one side has the dimension of width printed on the other side which also has a value of two for its two lines of direction.

The next coin in the sequence is three which has the number three on one side and line on the opposite side to represent the value of one point plus the two lines of direction of the line; and so on.

Now these coins can be used in a way similar to how the basketball was used to represent the geometry of ten- and eleven-dimensional space. And that is to provide solid-body objects to represent abstract mathematical values. Which, by the way, is also what the twenty-cent coin is used for.

So now you can select any one of these coins at random and perform the same trick that was done with the twenty-cent coin. You can flip it as many times as you like and each time it will land on one side or the other. So sometimes it will land on the decimal digit side and sometimes it will land on the spatial dimension side. But like the twenty-cent coin, regardless of which side these coins land on, the inherent mathematical value of each coin remains constant. So for example the number two coin can land either on the number two or on the dimension of width, but whichever side it lands on it will always maintain the value of two. And this same rule applies to all eleven coins.

I have established this particular model for a reason. And that reason is to challenge anyone who thinks that the mathematical proofs for the geometric structure of eleven-dimensional space are false.

Because in order to prove that this geometry is false, the only way you can do it is by proving that the eleven decimal digits of the decimal positional notation system are also false.

With the eleven coins, you can't just eliminate the value of one side of the coin without eliminating the value of the other side. If you destroy one side you destroy both sides. So the only way you're going to destroy eleven-dimensional geometry is by destroying the eleven decimal digits as well.

René Descartes.

René Descartes was a seventeenth century French philosopher and mathematician who is remembered for two important accomplishments: one in philosophy and one in mathematics.

In philosophy he set out to prove how things could be known to be true without question but ended up realizing that the only thing he could prove with absolute certainty was the fact that he exists as a thinking being: hence the term "I think therefore I am".

In mathematics he discovered analytical geometry which shows how geometric shapes can be translated into numerical values expressed as algebraic equations.

Back in 1987 when I established my first four facts, the third fact was the realization that the only thing I can prove to myself is the fact that I exist as a thinking being. So philosophically I had independently arrived at the same conclusion as Descartes.

Then in 2005, when I wrote down the mathematical proofs for eleven-dimensional geometry, I had to use the same type of analytical geometry that Descartes had used in order to establish those proofs, showing how geometric values could be translated into numerical values.

What this shows is that Descartes and myself were both clearly on the same wavelength, because we had both done exactly the same thing not just in one discipline, but in both disciplines of philosophy and mathematics. This also shows that there has to be some kind of direct link between the two disciplines because results like this don't just turn up accidentally or coincidentally. This 'link' will be explored in more detail in the final chapter.

Chapter 5

Space. The Final Frontier.

"The thing that often gets lost in the discourse is that string theorists follow the mathematics – they don't dream up this or that wizardry to explain some particular thing, they follow where the mathematics takes them."

Brian Greene

After I had established the mathematical proofs for eleven-dimensional geometry, any remaining doubts I had about their connection to string theory were totally annihilated. My assumption of 1997 was correct; the string theorists and myself had both used two completely separate mathematical processes that both ended up describing the same ten- and eleven-dimensional structures of space. The theorists had used the most complex mathematics on the planet, and I had used the simplest mathematics on the planet. And now there was no doubt in my mind whatsoever. By April 2005 I knew that the physicists and myself were both dealing with the same eleven dimensions of space.

Before I go any further there's something I need to clear up.

There are two theories that are being dealt with here. One of them is called string theory, which is ten-dimensional, and the other one is M-theory, which is the eleventh dimension.

Now sometimes these two theories need to be dealt with separately, and sometimes they need to be dealt with together as one complete, unified theory. But what I need right now is clarification.

It behooves me to have to do this but what I'm going to do, just for this chapter, is go out on a limb and invent a new word called 'strem' theory. Strem is a combination of the word string and the letter M. And it is the word that I will use when talking about the two theories as one unified theory.

So to clarify:

String theory is a theory of strings in ten dimensions.

M-theory is the eleventh dimension, and it is a membrane, not a string.

Strem theory is the two theories unified into one eleven-dimensional theory.

These two theories are both theories in quantum physics, and I've only got a high school education. So what would I know about theories in quantum physics? Not much, that's for sure.

But the one thing I do know about in great detail, and more so than anybody else on the planet, is the mathematical proofs for the geometric structure of eleven-dimensional space, and that's what matters. Because this geometry reveals a tremendous amount of knowledge about the space of our universe, much of which can be proven as facts of truth.

For example. Strem theory has used a series of complex mathematical systems to describe quantum structures that exist specifically in ten and eleven dimensions, and

I've used a completely separate mathematical system that describes the same ten and eleven dimensions.

Now you could argue that it's just a coincidence that both the physicists and myself end up with the same number of dimensions. But that would be foolish and naive. Why? Because strem theory is a well-established and clearly defined mathematical framework that is thoroughly understood by those who study it. And eleven-dimensional geometry is an irrefutable mathematical proof. They are two completely separate mathematical systems and they don't contradict one another.

On the contrary, they are both complementary to one another. The proof of one system confirms the proof of the other.

Physicists often say that the sign of a good theory is its ability to predict certain phenomena that can later be verified either by experiment or observation. This is precisely what strem theory has done by predicting a total of eleven dimensions of space which can now be verified by the mathematical proofs of eleven-dimensional geometry.

So now if you wanted to deny strem theory and prove that it isn't true, in order to do that you would also have to prove that the geometry of eleven-dimensional space isn't true.

And if you wanted to prove that eleven-dimensional geometry isn't true, the only way to do that is to prove that the mathematical values of the eleven decimal digits aren't true.

And if you wanted to prove that the eleven decimal digits aren't true, then you're a crackpot and you belong in the loony bin.

So unless you want to end up in the loony bin, you're really left with no choice other than to accept that string theory and M-theory aren't theories any more. Because they are now mathematically proven facts of the physical reality of the space of the universe.

String theorists use their mathematics to describe all the various properties of strings and membranes as they exist in higher dimensions at the Planck length.

What I can do is use eleven-dimensional geometry to describe the same strings and membranes from a different perspective.

And one of the first things I discovered was that this geometry is highly restrictive in what it allows. It literally dictates that if these strings and membranes exist the way that physicists describe them, then there's only one possible way they can do it, and there are no alternatives.

For example, if we go back to the basketball model, the basketball itself exists as a solid object in space. But what it is describing in string theory is a model of actual space as it would appear at the Planck length. And what it shows is that when space exists like this in ten dimensions, it is co-existing with nothing, which is represented by zero dimension.

So if space is co-existing with nothing then this must mean that this is where the space of the universe is actually created from nothing. And if this is so, then a string of energy existing in ten dimensions isn't a string that is existing in a ten-dimensional space. It's a string that is actually creating a ten-dimensional space.

So this is the mechanism that shows how the space of the universe first came into existence from nothing, just before it caused the big bang to happen. And it still retains all the characteristics attributed to it by string theory.

So long as it remains in ten dimensions, this string of energy is violently unstable because space and nothing cannot exist together once space has come into existence. Between the existence of space and the existence of nothing it has be either one or the other; it can't be both.

So it is immediately forced to rupture and split in order to separate the six external dimensions from the four internal dimensions and keep them permanently separated by forming a membrane of space which becomes an eleven-dimensional point particle of space- energy, with the six external dimensions fixed permanently on the outside and the four internal dimensions fixed permanently on the inside.

But an eleven-dimensional membrane by itself cannot expand into the four internal dimensions. It remains permanently fixed at the Planck length in order to maintain the separation of the two groups. But it has to sustain its ability to keep nothing from existing.

The only way it can do that is by initiating a chain reaction around the periphery of the membrane, causing more strings to form and collapse into membranes and allowing for space to start expanding into four-dimensional spacetime.

The initial stages of this expansion were so swift and violent that they became known as the big bang. And the expansion of space that they initiated still continues to this day.

The idea of space expanding from tiny points at the Planck length into the vastness of the universe we see around us today seems like a bit of a stretch. But it becomes a bit more realistic if we remember that something that is expanding at the Planck length is also expanding at the Planck time, which is 10^{-43} seconds, or one ten millionth of a billionth of a billionth of a billionth of a billionth of a second. So let's say a point of space doubles in size each time, this is how many times it has doubled in a single second.

And if you think that sounds ridiculous, well maybe it does. But it's the result I end up with from applying eleven-dimensional geometry to the dimensions of strem theory.

Internal point is the overall eleventh dimension. But from our position on the inside of the universe it's the fifth internal dimension; the point from which the four linear dimensions expand into the four-dimensional spacetime that we are now living in. It is the smallest and most fundamental particle in the universe: the particle from which all other particles of matter and energy have evolved, and that includes us.

So what this ultimately means is that space isn't just the stuff that we exist in, it's also the stuff that we're made of.

Dark Energy.

It's estimated that dark energy represents about 70% of the known energy of the universe. And what it does is it causes the expansion rate of the universe to accelerate. It's hard to know exactly when this acceleration started. One estimate puts it at nine billion years ago.

Eleven-dimensional geometry can be used to explain dark energy. Though whether it's right or not remains to be seen. But it's like everything else with this geometry. It simply states that if dark energy exists, then this is the only possible way it can exist and that's it.

When we think about the expansion of the universe, a common analogy to use is an inflating balloon. As the balloon expands its surface spreads out evenly in all directions. And on the macrocosmic level that's how space appears.

But down at the Planck scale it's a different picture altogether.

If we could see space expanding at this scale what we would be looking at is something visually similar to a pyroclastic flow as it rolls down the side of a volcano. The main difference is that as a pyroclastic flow spreads out and settles it loses its energy, whereas space is doing the opposite.

Space at the Planck length is actually creating new energy. So if the expansion rate of this energy starts to increase as the periphery regions of the universe continue to grow, this increase in the energy would then cause the expansion rate of the universe to accelerate, thus providing the source of dark energy.

Eleven-Dimensional Supergravity.

This is a theory closely related to M-theory, and I know absolutely nothing about this theory beyond its title. So with my educational background I really shouldn't be messing around with theories like this.

But the fact is that eleven-dimensional geometry provides a possible explanation for the source of gravity in the eleventh dimension.

Whether this explanation is viable or just completely off the wall I have no way of knowing. So I'll just put it out there and let physicists decide if there's anything to it or not.

The eleventh dimension has two crucial roles to play in order to maintain a steady, expanding four-dimensional universe.

Firstly, it has to keep those six pesky external dimensions permanently locked on the outside. And secondly, it has to keep four-dimensional spacetime together as one coherent and continuous body.

So in order to maintain these two functions it would have to stay in a permanent state of contraction or tension: like a rubber band that's been stretched. It's always trying to pull itself in. So it would be this contraction, occurring at every point in space at the Planck length, that would be the fundamental source of universal gravity.

Dark Matter.

Dark matter is the name given to regions of the universe where the gravitational effect is greater than can be accounted for by visible matter only.

So, what if dark matter wasn't matter at all? What if it was just heavy space? Regions of space where the accumulation of eleven-dimensional supergravity builds under pressure, much like the water pressure in the oceans build as the water gets deeper.

That's the best explanation I can give for dark matter.

Multiverse.

Multiverse is the idea that there is more than one universe.

There isn't.

Zero dimension forbids it.

Unification.

As a theory of everything, one of the aims of strem theory is unification. Like for example the unification of the four energy forces into one single force.

What you get with the eleventh dimension is the unification of mathematics and physics into one single entity.

In mathematics, the eleventh dimension is called 'internal point', and it is a geometric unit of measurement.

In physics the same dimension exists as a point particle of space-energy at the Planck length.

Fundamentals.

The eleventh dimension is the most fundamental particle in the universe, from which all other particles of matter and energy have evolved.

What eleven-dimensional geometry shows is that in order to arrive at the mathematical proofs for the eleventh dimension - the most fundamental particle in the universe - you have to be using the most fundamental mathematics in the universe to do it, i.e. the eleven geometric dimensions and the eleven decimal digits. That's what makes it so easy to understand.

The Laboratory of The Mind.

One of the most compelling arguments against strem theory is that it is a purely mathematical framework that can never be tested in a laboratory or by any other physical means. The reason for this is because at the Planck scale the sizes are way too small and the energies way too high for any of the technology or equipment that we have been able to produce or are ever likely to produce.

In fact, getting a physical look at eleven-dimensional space is fundamentally impossible because it means we would have to invent a machine that could exist on the outside of the universe where there is nothing for it to exist in. It's just never going to happen.

But the mathematics have shown that there is a clear and solid link between string theory, M-theory, eleven-dimensional geometry and the eleven decimal digits.

This link can be clearly understood without the necessity of having to know the precise mathematical details of the two theories. You only need to have a general understanding of what the theories propose; just like I do.

So what this means is that rather than using a regular laboratory, full of Bunsen burners, microscopes and test tubes, we use instead a very special type of laboratory known as the Laboratory of the Mind. This is the laboratory we can use to test those mathematical links in order to prove the results for ourselves.

And basically anyone who has the ability to count to ten has already got the mathematical qualifications to test those links in their own Laboratory.

Furthermore, anyone who can count to ten pretty much accounts for every single member of the human race, which currently stands at a population of over eight billion people. So that's over eight billion individual Laboratories, each capable of independently testing the viability of these mathematical proofs. I think that should be enough to decide whether it's right or not.

Another thing physicists like to say is that a good solution to a theory is one that can be understood by anyone rather than just a small group of highly educated academics.

And that's precisely what we have here, isn't it.

Chapter 6

<u>Towards The End of The Quest.</u>

<u>Conclusion.</u>

In chapter four I mentioned that there clearly must be a connection between the philosophy and the mathematics. And that was because René Descartes and myself had both done essentially the same thing - twice. I will now attempt to explain what that connection is and why it is necessary.

When I set out in 1987 at the beginning of this quest, it was a quest to determine the truth about the reality of my existence in order to find a reason for why I existed. But the truth that I was searching for wasn't my version of the truth, or your version of the truth, or their version of the truth, or even God's version of the truth. No. The truth that I was searching for was the actual truth, whatever that truth may be. And I wanted it to be based on facts that I could know with absolute certainty to be facts that I could know and prove to myself to be true.

So let's review those first four facts that I established.

They were:

Fact 1: I know for a fact that I can trust any naturally occurring pattern.

Fact 2: I know for a fact that I cannot trust my own belief system.

Fact 3: I know for a fact that I exist as a physical form of conscious intelligence.

Fact 4: I know for a fact that I exist in space.

Those four facts between them led me directly to the discovery of two completely separate geometric patterns that were both showing me the exact same result: they were both showing me the internal point of view and the external point of view both together at the same time.

Eighteen years later it was those two patterns, together with string theory and M-theory, that led me to the development of the mathematical proofs for the geometric structure of eleven-dimensional space. And they are irrefutable mathematical proofs, which means eleven-dimensional space is a mathematically proven fact of the physical space of the universe, not merely my opinion.

And remember, if you choose to disagree with that, then it's the loony bin for you.

So what I've learned is that it's space itself that is the actual cause of everything that exists in the universe. It is also the most fundamental particle of everything that exists, including us.

Not only that, but in order for us to discover this reality we have to be using higher-dimensional geometry in order to be able to step outside of the universe and view the space of the universe from an external point of view.

This is a remarkable achievement, that we can now use our mathematical knowledge to separate ourselves from the universe that we exist in in order to see the mechanism that initiated the universe's beginning by being able to look at it from the outside.

So getting back to the connection between the philosophy and the mathematics, it is this. In order for us to get to the truth about the reason for our existence, we have to first get to the truth about the origins of the universe. And the only way to do that is through the knowledge of the mathematical proofs for the geometric structure of eleven-dimensional space.

Right from the outset, as soon as space first came into existence, the first thing it needed in order to exist was ten-dimensional geometry. Because without that geometry there would be no structure.

And as soon as that geometry exists, it's possible to derive from it the inherent and individual values of the ten decimal digits.

So before space has even collapsed into the eleventh dimension, it has already established the basic foundations of all mathematics.

Then, after billions of years of the evolution of the space of the universe, followed by billions of years of the evolution of life on planet Earth, we end up developing the ability to prove mathematically, with irrefutable mathematical proofs, that this is the geometry that caused the universe to come into existence in the first place.

And following the typical characteristics of this geometry, this leads me to the only logical conclusion that makes any sense regarding the question of our existence.

Right from the outset, as soon as space first came into existence, those five groups of ten-dimensional strings mentioned in the preface didn't have any choice in the matter. From the moment they first formed, they were driven to organize themselves in such a way as to ensure the creation of a universe that allows for the development of conscious intelligence, because that is precisely what has happened on this planet, and we are it.

We are the Conscious Intelligence of the Universe.

That is what gives our lives meaning and purpose.

It is the reason why we exist.

The End.